A SIMPLE LOGICAL KEY TO EXPLAIN OUR PHYSICAL UNIVERSE

Michael L. MacLaughlin

A SIMPLE LOGICAL KEY TO EXPLAIN OUR PHYSICAL UNIVERSE

First published in Australia by Michael L. MacLaughlin 2017

Copyright © Michael L. MacLaughlin 2017
All Rights Reserved

National Library of Australia Cataloguing-in-Publication entry:

Creator:	MacLaughlin, Michael L., author
Title:	A SIMPLE LOGICAL KEY TO EXPLAIN OUR PHYSICAL UNIVERSE
ISBN:	978-0-6481328-0-6 (pbk)
Subjects:	SCIENCE / Physics / General
	SCIENCE / Physics / Quantum Theory
	SCIENCE / Cosmology

Also available as an ebook: 978-0-6481328-1-3 (ebk)

Typesetting and design by Publicious Book Publishing
Published in collaboration with Publicious Book Publishing
www.publicious.com.au

To all of those who dare to daydream
while being taught what to think

Introduction

If you are not a physicist but want to feel comfortable in your own mind that you understand the universe you live in, then this little book is for you. I will give you a key, a simple logical view of a single particle that you can use to explain all of the physical properties of our universe. The key is based on five rules that all magnets must follow. Once you have an understanding of this key you will be able to use it as a microscope for your mind's eye to look at and understand everything that is too small to see, such as the minute singularity that was our universe before the Big Bang. You will see how electrons, protons and neutrons, the building blocks of atoms, were created during the Big Bang and how they go on to form the elements on the periodic table. You will understand precisely what forces were at play during the whole process and how they now continue to shape the way the universe and everything in it behaves.

This key also acts as a telescope for your mind's eye that will let you see how planets, solar systems, galaxies and black holes were formed and how they continue to interact with each other. With this mind telescope you will be able to visualise the dark energy force that is causing the expansion of the universe and also understand the process that makes us think dark matter exists. You will know why it is safe to predict that all planets in all solar systems in all galaxies in the whole universe have an abundance of life giving water, carbon and oxygen, and how it is indeed the dark matter effect that makes it so. You will know why the magnetic poles of all quasars, the remnants of entire galaxies that have been consumed by their black holes, all happen to line up in the same direction, even though they are separated by billions of light years of space and time.

You are not going to study quantum or particle physics or even look at a single mathematical equation, but with the simple logical view I will give you, you will understand the two great mysteries contained within Paul Dirac's equations that form the basis of our mathematical understanding of the quantum realm. Namely you will know precisely why it is that they require the universe to be filled with a sea of electrons and also why his equations predict that there is an equal amount of anti-matter, somewhere out there in the universe, to balance out our matter. You will also come to realise that this is what the magnetic poles of all of our quasars are pointing to.

You are not going to need to get into a tangle trying to understand all of the different and often contradictory forces that shape our universe. In fact, you need not even be aware that they all exist because with this simple logical view I will give you, they will seem so blindingly obvious. You will know what causes an electron to behave so strangely. You will know what holds atoms together to form molecules. You will know how and why gravity comes into being when an electron is separated from a proton to form a neutron, and how this same concept gives some materials their magnetic properties. You will understand how these particles were created when things travelled faster than the speed of light during the Big Bang, and you will see that it is this same Big Bang energy that is released during nuclear explosions. There should be no mystery in the physical universe that you can't solve with this simple key that I will give you in these following pages.

1. The Key

The key is based on five rules that all magnets must follow. This picture of iron filings sprinkled onto a piece of card that covers a bar magnet will help you better understand these rules.

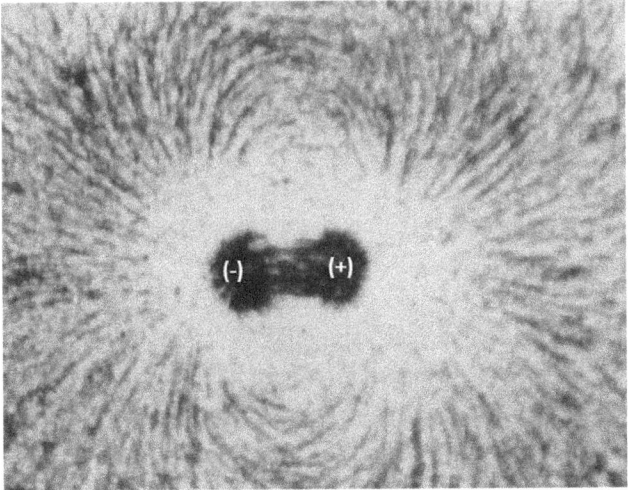

These rules are:

1. A magnet emits charged magnetic fields at both of its two poles that are equal in strength but opposite in charge. One pole field we call positive and the other negative.

2. When magnetic fields encounter others of their kind, the oppositely charged fields attract each

other, and the identically charged fields repel each other. So a negatively charged field will attract a positively charged field but repel another negatively charged field.

3. The fields emitted by the magnet obey the inverse square law. This means that with each step you take toward the magnet, you feel the force of its field increase by the square of the distance you have travelled. Simply put, if you take 2 steps closer, the force increases by 2 times 2 equals 4, if you take 3 steps closer the force increases by 3 times 3 equals 9, and if you take 4 steps closer the force increases by 4 times 4 equals 16. Going in the opposite direction away from the magnet, 2 steps decreases the force by 4, 3 steps by 9, and 4 steps by 16. This means that from a distance you don't even know the force of the magnetic field is there, but as you start to get very close it gets dramatically stronger. This is what you feel when you try to force two magnets together by their identically charged poles. As you get very close it is almost impossible to hold them together. The opposite occurs for oppositely charged poles, where they will hang on to each other desperately when they are joined together but they rapidly lose interest as soon as you manage to pull them apart. This is an important concept to grasp because all

the forces in the universe obey this same inverse square law, from the very weak but very large gravity to the very small but very powerful atom.

4. Magnetic fields cannot travel faster than the speed of light (approximately 300,000 kilometres per second). This means that if it were possible to move a magnet faster than the speed of light, its two charged magnetic fields would be left behind. Generally, you don't have to worry about the magnetic field being left behind when you move a magnet – but if you could move it instantly from the Earth to the sun, the magnet's fields would only catch up eight minutes later. There was once a brief moment in time when things did move faster than the speed of light, known as the inflationary period of the Big Bang. Bear this in mind because as you will soon see, it has some surprising consequences.

5. When two magnets are joined together by their opposite poles, the combined magnetic fields they generate are more powerful than those of each single magnet. The field of one magnet passes effortlessly through the body of the other magnet and combines with that other magnet's field as though they were one. Well almost, we must always remember rule 3, the inverse square law where the strength of the magnetic field being

generated decreases with distance from its original source magnet.

Now that you know the rules for applying the simple key, you are going to have to use your imagination, because magnetic fields can't be seen with the naked eye but only inferred by their action on their surroundings. At a quantum level, things get too small and too powerful to be seen by any method other than the imagination. Don't be shy to use your imagination for it was Einstein who said "Imagination is more important than knowledge. For knowledge is limited to all that we now know and understand, while imagination embraces the entire universe, and all there ever will be to know and understand."

Now imagine very small but extremely powerful magnets. They are so small that three of them fit together inside the nucleus of a single proton which makes up the core of the smallest atom, hydrogen. To align our simple key and our thinking with that of particle scientists, we will call these extremely small and extremely powerful magnets 'quarks'. Now let your imagination take all of these quarks – from all of the atoms in all of the matter in all of the solar systems, in all of the galaxies, in the whole universe – back in time to moments before the Big Bang. That is easy; you end up with this same picture you saw before.

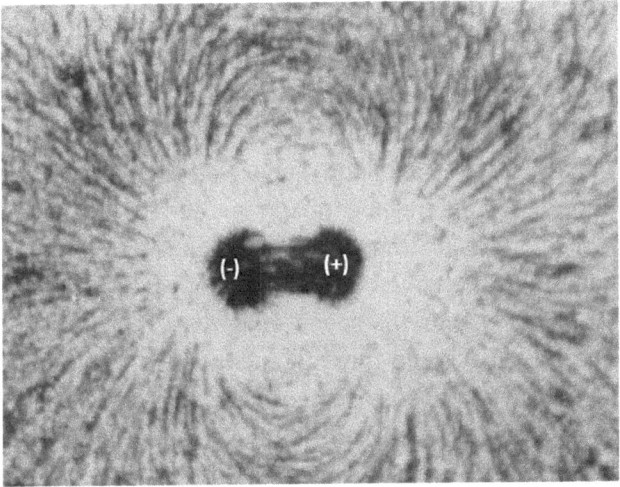

All the quarks in the universe would be tightly bound together to make up a single super-massive magnet we will call the 'singularity'. Each magnetic quark particle would be contributing to the magnetic strength of the whole. The singular magnet would generate two very large, very empty fields that stretch to the ends of the universe, one side negatively charged and the other positively charged. Scientists tell us that a magnet can only grow to a certain size and strength after which it will explode, and that this size is many orders of magnitude greater in strength than the largest known black hole in the universe. This is the point at which our single massive magnet, the singularity, will have gathered enough quarks and enough strength to cause it to explode, in what we know as the Big Bang. Scientists have also calculated that there must have been a brief period, moments after the start of the Big Bang, when the universe expanded much faster than the speed of light, known as the inflationary period. As the fleeing quarks exceeded the speed of light they left the magnetic fields they had been generating behind, because magnetic fields cannot travel faster than the speed of light. At this point the quarks were forced through and into either one side or the other of those same universal magnetic fields that they had been contributing to. I say 'through', because the universal magnetic fields are inversely squared in strength, being significantly denser at the initial site of the explosion and rapidly weakening as the quarks passed through the densest part and headed away. As the quarks were blasted in all directions and overtook either one of their magnetic fields, depending on which side of the universe they happened to come out on, something weird happened. Again, to help your imagination I offer this following picture of iron filings scattered over a card that is placed over an arrangement of magnets.

A single magnet at the centre represents a quark that is trapped within one side of the universe's two magnetic fields. In this picture, the one side of the universal magnetic field is being generated by gluing a circle of magnets around the quark, with all of their

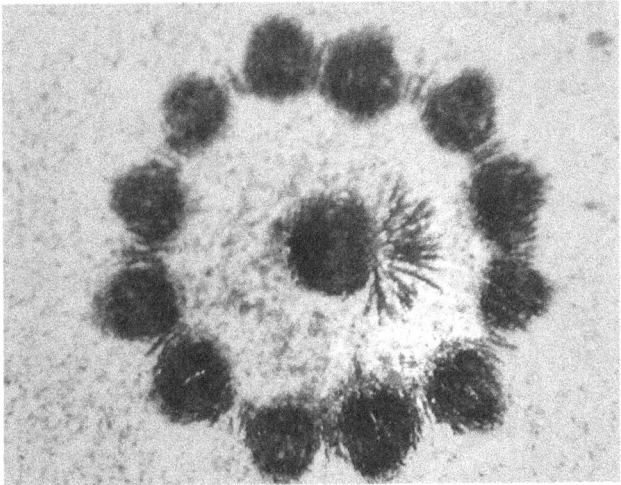

similar magnetic pole fields pointing in towards the centre. Look at how the quark is behaving within the imaginary mono-polar universal field we have created. Its oppositely charged magnetic field is being attracted and stretched by the surrounding universal field, while at the same time its similarly charged field is being repelled and bunched up by the same universal field. Stretched and weakened on the right-hand side, and bunched up and strengthened on the left. When you live on the same side of the universe with this quark, you easily notice its 'bunched up repelling side' but really struggle to find evidence of its 'weaker stretched side'. This quark in this guise is the only type of particle we need to build the whole universe and explain everything in it and, along with our five rules of magnetism, forms the basis of our simple key.

I will show you in the following pages how we can use this simple key with its single particle to explain many of the really big mysteries currently facing our greatest scientists. Along the way I will try to give you a much better understanding of how our simple key with its single type of particle makes everything possible. We will start at the very small end of the scale and explain how atoms are made and work our way up to the vastness of disc galaxies with their central black holes, and all the while connect the dots to show how our tiny quark particle is responsible for everything in the universe. I aim to show you also how the humble quark is responsible for creating all the forces that we know of in the universe, from the Big Bang to the tiny electron with gravity, dark energy and atomic bonds thrown in along the way for good measure. Hopefully by the end of the book there won't be anything about the physical universe that you can't easily understand and explain with our simple key.

2. Particle Physics

Most ordinary folk who like to piece things together logically in their mind will have found this area of science the most difficult tangle of illogical ideas and rules that just can't possibly make sense. But if we apply our simple key you will find that in fact it is all incredibly simple and perfectly logical. Let us take another look at our photograph of the universe before the Big Bang.

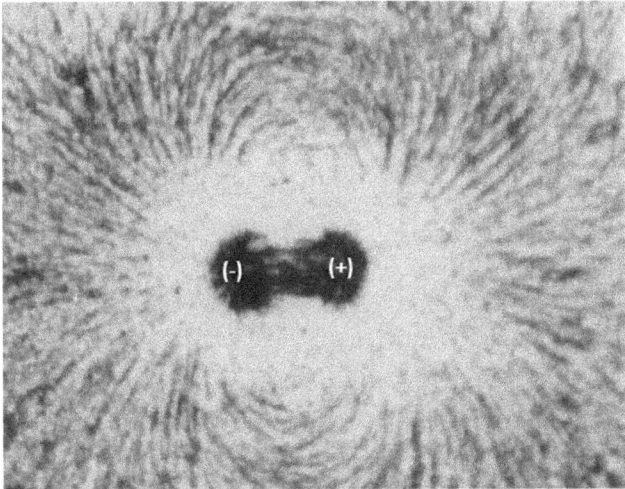

I mentioned in my introduction that the genius Paul Dirac developed equations that form the basis of our mathematical understanding of quantum mechanics and particle physics. These equations posit two great scientific mysteries; first that the universe must be filled with a sea of electrons and second, that there must be an equal amount of anti-matter to balance the matter in the universe. We know from our simple key that there are two universal seas, not of electrons but rather of charged magnetic fields. Dirac's sea of electrons is in fact the sea of negative magnetic energy that exists on our side of the universe from before the Big Bang (left-hand side of the photograph). We also know that the other half of the universe (right-hand side of the photograph) is filled with a sea of positively charged magnetic energy in which everything appears to be the same, but where all the polarities of the charges are reversed, making it magnetically and electrically entirely the opposite. This is the anti-matter side of the universe required by Dirac's equations. I will explain later how recent astrological observations might indicate where the boundary is between the matter and the anti-matter halves of the universe, but for now let us relish that fact that using our simple key we know why Dirac's equations need both the sea of electrons and the existence of anti-matter to be able to explain how particles behave at the quantum level.

When using our simple key to explain particle physics, we need to be aware that there has always been a great controversy over what defines a particle.

Remember, down at this level it is impossible to actually see anything, so scientists have had to imagine possible particles to explain what they are seeing. Particle physicists describe the inside of atoms as electrons remotely circulating around a central nucleus. The nucleus consists of protons and neutrons, and where the protons are positively charged, the neutrons have no charge. The protons and neutrons are made up of yet smaller particles called quarks, each proton and neutron having three quarks at their core. They cannot find anything smaller than the electron and the quark and so believe these two particles to be the basic building blocks of our universe.

Our key is based on our understanding of the quark, but what of the electron? This so-called particle is the one that started the great controversy about what properties are required by something in order for it to achieve 'particle' status. When examined, the electron appears to have some permanent particle-like qualities yet at the same time exhibits wave-like properties. Many great scientists, including Einstein, were very unhappy when, in the early 1900s the majority of eminent particle scientists declared that the electron was indeed a particle. This has allowed for another twenty or more particle-like phenomena since discovered to also be awarded particle status, even though none of them exists in space and time on their own in any form for more than a single second. For the purposes of our simple key we can ignore these, but we will explain exactly what the electron is and why it behaves the way it does.

To begin, we need to look again at our photograph of the quark.

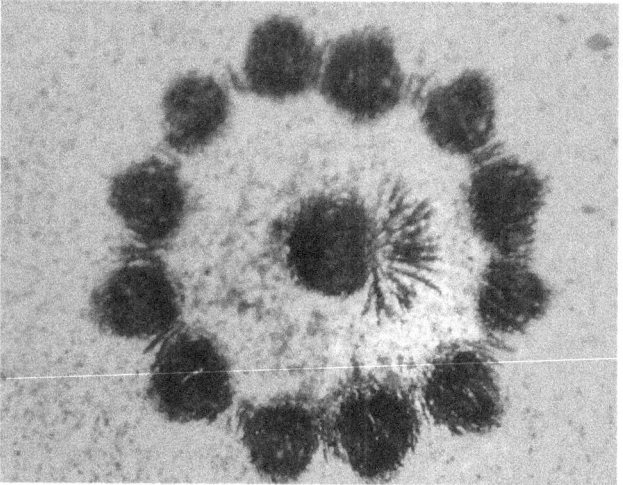

Quarks combined in space after the Big Bang because they have both negative and positive magnetic poles, even though the strength of either of their poles was made uneven by the background universal magnetic field that the inflationary period forced them into. It is only after three quarks have combined to form a proton on our side of the universe that they repel other quarks. At this point, they together have three large negatively charged poles that point in three different directions, being bound by three weaker positive poles. The quarks larger negative poles repel each other internally to form an all-encompassing negatively charged cloud. This cloud then repels the surrounding negatively charged universal field and also any other protons that try to approach. When scientists examine a proton, therefore, it appears to have a surrounding negatively charged electron that is moving so fast that it is able to form an electron cloud that then repels anything that tries to approach it. Thinking about our simple key, we can see how our three quarks that make a proton and electron will behave precisely the way science tells us they should. Our simple key tells us that an electron is not an actual particle, but rather a bunched-up cloud of negatively charged magnetic field, surrounding a proton that is being repelled by the surrounding negatively charged universal field. It also explains the annoying ability of the electron to be everywhere at the same time. This idea fits perfectly with science's mathematical interpretation of particle physics.

If we use a lot of energy and a large moving magnetic field, it is possible to dislodge a proton's electron to the extent that the proton ends up with a magnetic field that has no charge (when compared to the surrounding negatively charged space field).

In this altered state, the proton becomes a neutron and does not repel anything. Our simple key tells us that the surrounding bunched-up negatively charged magnetic field is the electron. The bunched-up negatively charged field that has been dislodged from the proton becomes magnetically bottled up by the surrounding negatively charged repelling universal field. A bottled blob of magnetic field, that is, that can travel through space as a wave but arrive as a particle, behaving precisely the way science tells us that an electron behaves. Science tells us that if left alone and uninhibited, the neutron regathers its lost electron from space over a period of approximately 15 minutes, at which point it turns back into a proton. This is exactly what you would expect from our simple key model.

What I have described here is what we would expect if we think about our simple key, the five rules of magnetism, and our two photographs – and mirrors precisely what science tells us about the behaviour of protons, neutrons and electrons. In fact, with this key, we can explain any of the science surrounding protons, neutron and electrons. The model fits perfectly. Science on the other hand has no way to explain its own mathematical starting point of having a sea of electrons and the existence of anti-matter. Our simple key tells us how quarks were given their properties by being blasted into the negative side of the universal field, and so formed into protons that can be turned into neutrons when their electron cloud is removed. The same process happened on the opposite side of the universe, only there the universal field is positively charged and the polarity is reversed to make anti-matter from anti-protons, anti-neutrons and anti-electrons.

If you explore the other twenty or more so-called particles discovered by science you will see that they have been manufactured by bashing protons and neutrons together at great speeds. The resulting electromagnetic waves have been classified as particles in the same way as electrons were. These misnamed particles are simply electromagnetic waves produced in the universal magnetic field. The most glaring example of this folly is the photon, the so-called particle that carries light. Our simple key gives a much more logical explanation than light being carried by a particle. When different shaped molecules are energised and excited, as in the electric light bulb – where electrons are forced through a filament of magnetically bound metallic particles – they spin and, because of their shape, alternately push and pull on the universal magnetic field, causing waves to be generated in the same way you can generate waves in a swimming pool by pushing and pulling yourself against the side of the pool. The size and shape of the spinning molecule determines the length of the wave that is produced and so the colour of light that is being emitted. We don't need 'photon particles' to carry light, and then have to worry about why they are travelling through space as waves. We only need there to be a universal magnetic field, the same one required by Dirac's sea of electrons for his mathematics. In fact, we don't need any more particles than the single fundamental quark described in our simple key to be able to make and logically explain everything in the universe, so let us move on from particle physics to larger things.

3. Atoms, Gravity and Magnets

If we take a look at the periodic table of elements we see that the smallest element is hydrogen and is said to consist of one proton with its electron. Thinking about the Big Bang and how individual quarks were blasted into the universal magnetic field, it is no surprise that space appears to be filled mainly with hydrogen atoms and that scientists estimate that hydrogen makes up 75% of all of the matter in the universe. Think back to our description of how quarks formed protons, it is no surprise that hydrogen atoms don't like each other and try to keep as far away from each other as possible on account of their bunched up repelling negatively charged magnetic fields and is also why hydrogen happens to be a very light gas.

It is possible to crunch protons (hydrogen atoms) together under extreme circumstances, such as those found in burning stars and nuclear explosions. When this happens, electrons are expelled due to the roughly circular shape of protons – it is a physical and mathematical requirement of circular objects that their surface area decreases in relation to their volume of contents as their size increases. Thus, the more protons you squeeze into an atom, the less of their electrons it can hold on its surface. When those protons lose their electrons, they become neutrons. This is why smaller atoms shed electrons to form larger atoms and why larger atoms contain both protons and neutrons. Once contained within an atom, the force a neutron can exert to reclaim its lost electron field is diminished by distance, according to the inverse square law. The further it gets away from the outside surface of the atom, the less force its positive pole can bring to bear on pulling in the surrounding negatively charged universal field. The positive pole does not switch off, it just gets too far away from the universal field to be able to regather its lost electron as it is buried under all the other surrounding protons and neutrons. The electrons that are able to be pulled onto the surface of a larger atom ensure that any other atoms in their vicinity are repelled by their negatively charged electron clouds. Remember, when we talk about electrons we are never talking about actual particles because they do not exist, but rather an equivalent amount of negatively charged magnetic field that is being bunched up against the negative pole of our magnetic quarks. The protons that have lost their electron clouds to become neutrons now exert that electron gathering force at a distance as gravity. As an element on the periodic table gains more neutrons and so loses more electrons it becomes heavier. That heaviness is the pull of neutrons trying to regather their lost electrons and is expressed as gravity at the quantum level.

A hydrogen atom has only one proton and no neutrons, and so generates no gravity. The next element on the periodic table, Deuterium – also known as heavy hydrogen – has one proton and one neutron and so is the first atom to exert any gravity. An atom of lead has 82 protons and about 126 neutrons and so exerts a lot of gravity and is said to be a heavy element. If you use our key to look at the periodic table you can see why different elements have different physical properties based on the number of protons and neutrons at their core. The more 'electrons in their outer shell', the more likely they are to be gasses, because they have more protons than neutrons and so carry a greater negative magnetic charge to repel other atoms. Those with only a few 'electrons in their outer shell' have more neutrons than protons and so exert more gravity to pull other atoms towards themselves and are more likely to be solids. Very large elements become unstable and decay radioactively emitting nuclear radiation in the form of neutrons. Neutrons have no negative repelling force and pass effortlessly through the universal field and into the nuclei of other atoms, sometimes altering them. Adding or subtracting protons and neutrons from the nucleus of an atom changes its position within the periodic table of elements, and so its chemical properties, making fast moving wayward neutrons dangerous to life. When small elements are fused together to form larger ones under the extreme gravity conditions found in burning stars they emit electrons and thus form heavier atoms with more neutrons, increasing the stars gravity and further fuelling the process. When you keep our simple key in mind, all of these previously difficult to comprehend concepts suddenly make perfect sense.

Science tells us that magnetic materials get their magnetic effect from the direction in which their electrons are pointing. Materials that have no magnetic field have electrons pointing in random directions, which cancels out any net magnetic effect. Permanent magnets have electrons that all point roughly in the same direction and are held there by the latticed arrangement of their atoms. Induced magnets are materials whose electrons line up when they come under the influence of a nearby magnetic field. We can say electrons are responsible for generating magnetic fields, while the lack of them generates gravity. Let us look again at our photograph of the quark.

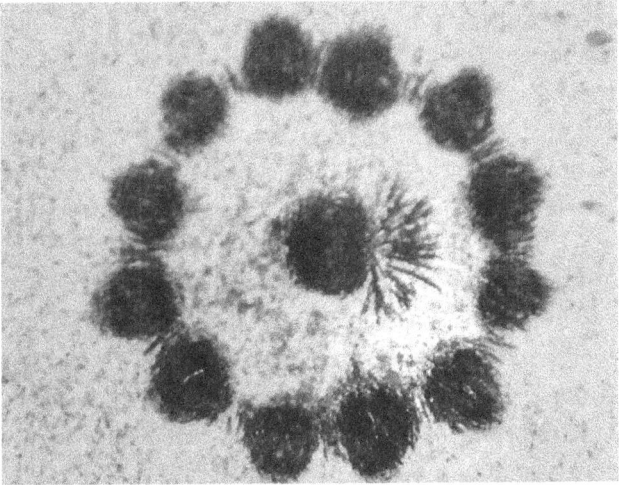

Imagine that the two forces you are looking at here are the electron on the bunched-up, repelled left-hand side, and gravity stretching space on the right-hand side. Gravity and the electron appear to be incredibly strong if we think about the Earth hanging onto the moon or the power of the Sun's rays as it fuses atoms and expels electrons, but they are nothing compared to the strength of the magnetic fields of each quark, inversely squared of course. We can't comprehend the strength of the quark's magnetic field because three of them are always bound tightly together, each pole cancelling out the effect of their neighbour's opposite pole. What we are seeing in our electrons, gravity and magnets is only the effect of the quarks having been blasted through the light speed barrier and now residing within the magnetic field they had generated while they were once part of the singularity. We must not forget the other side of the universe, where polarities are reversed, a place where Dirac's equations require an equal amount of anti-electrons to balance out electrons. No doubt there will also be an equal amount of anti-gravity to balance out our gravity.

To summarise, we can say that three quarks bind together to form a proton which becomes a neutron when it loses its electron cloud. Once it loses its electron cloud, the neutron tries to pull it back in from space and this pulling force is gravity. The amount of gravity exerted by planets and stars and galaxies depends on how many neutrons they have in all of their atoms. This tug on space creates what scientists describe as a gravitational hole in space-time around large objects that smaller objects fall into, as is our moon falling into Earth's gravitational hole but is kept out by the speed at which it is travelling.

4. Dark Energy

Something is causing everything in the universe to move away from everything else at ever increasing speeds. More distant galaxies are now moving away from us at speeds greater than the speed of light which means we will never see them again. Science has labelled this unknown pushing force as Dark Energy. To explain this dark energy force, we need to begin at the beginning, just before the Big Bang. To visualise the shape of the magnetic fields that existed at that time, let us look again at our photograph of the universe at that moment.

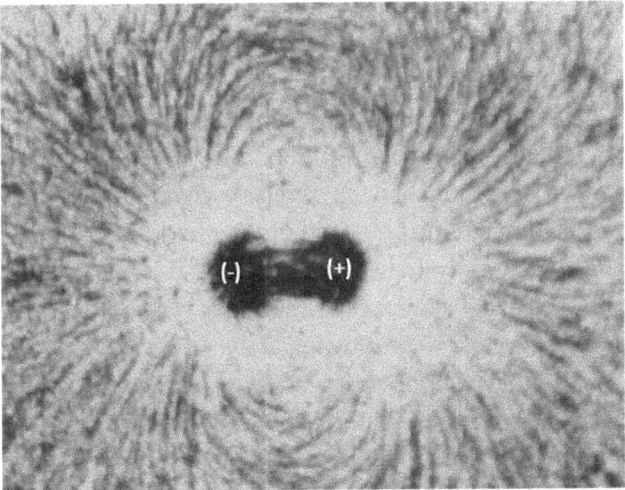

The density or strength of each of the universal magnetic fields was very great close to the site of the singular universal magnet at the time of the Big Bang, and weakened or thinned out as given by the inverse square law with distance from the centre. If we think again about our quarks, as they exceeded the speed of light they were blasted into and through the densest part of the fields, closest to the centre. Once on the other side of the densest area of magnetic field, they would be constantly repelled away from this area and towards the ever-increasingly less dense field in the distance. As their dominant magnetic field is the negatively charged electron field, it will be repelled by the magnetically sloped negatively charged universal field, and will always be moving from the densest area towards the less dense area, like a skateboarder rolling down a slope. The slope is created by the inversely squared field density, and the force pulling the skateboarder down the slope is the large bunched up similarly charged and thus repelled magnetic pole field of the quark.

The force will always be there and is the reason that the expansion of the universe is accelerating and neither constant or slowing down. There is always that gentle push down the slope, making everything expand faster and faster. Well, that is half the story but there is another force at work and that is the fusing of atoms in stars, and the resulting expulsion of electrons to make neutrons and heavier elements and more gravity. The expelled electrons are balls or waves of magnetic field energy with the same electrical charge as the universal magnetic field the stars are burning in. This expulsion of electrons has the effect of increasing the field density around all burning stars, and so increases the dark energy pushing effect between all the matter in the universe. Remember, magnetic fields cannot travel faster than the speed of light, which makes this process at a galactic level seem very gradual. Dark energy is called by that name because the force that is causing the accelerated expansion of the universe is mysterious to science, but as you can see is relatively easily explained with our simple key.

5. Dark Matter and Black Holes

While on the subject of dark and mysterious things, let us use our simple key to explain dark matter. When scientists examine the way stars, planets and other materials rotate around black holes at the centres of all disc galaxies they notice something odd. If all of this matter in the disc is being pulled in towards the black hole under gravity, the material closest to the black hole should be moving much faster than that which is much further away on the outer edges of the disc. But that is not the case. The material on the rim is travelling faster through space than the material near the centre, and their best explanation for this phenomenon is that there must be a lot more matter generating a lot more gravity in the disc than we can see with our telescopes. We can't see it so we call it 'dark matter'. Our simple key gives us a much more obvious explanation for what is happening, but before we can explain dark matter we must first examine black holes.

Science tells us that black holes form when very large stars reach the end of their lives, have burned all the atoms in their cores and expelled all of their energy, leaving only their enormous gravity. Something that puzzles scientists, however, is that this process should take many billions of years and yet, if we look back in time and space, we find that some black holes appeared shortly after the Big Bang. Using our simple key, we can see an easy explanation for black holes and their early formation. Put simply, a black hole is a pure magnet with many quarks tightly bound at its core. It is no longer much influenced by the universe's background negative magnetic field, as its pure magnetic field generated by the tightly bound quarks and governed by the inverse square law is locally much stronger. Any more than three quarks, and the three-dimensional proton/electron repelling shape changes to one of a pure magnet with the strength to pull in any nearby protons, strip them of their electrons, and accumulate their quarks in an ever-growing mini black hole. During the Big Bang not all quarks would have been ejected as individuals and there would have been many of these fragments of various sizes, containing many tightly bound quarks ejected as already formed black holes. To summarise, our simple key tells us that all the quarks in the universe formed a giant magnet, and those larger fragments of the magnet ejected during the Big Bang are the black holes that formed so early on in the evolution of the universe. In the beginning there was only a single massive black hole that exploded causing the Big Bang. Some of the energy of the Big Bang became trapped as electrons creating matter when it passed through the light-speed barrier during the inflationary period. Burning stars are releasing this Big Bang energy and when the process is complete they revert back to their original black hole state.

Our simple key tells us that a quark is a magnet and that black holes are very large magnets, but what does it tell us about the dark matter that causes the rim of disc galaxies to revolve at such high speeds? The key to understanding dark matter is in knowing how material accumulates to form the discs of galaxies.

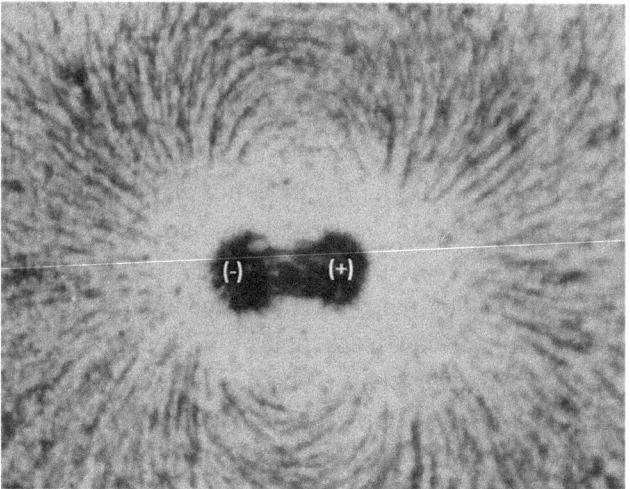

Let us take another look at our photograph of a bar magnet. Imagine a vertical line down the centre of our photograph between the positive and negative poles. This line has no magnetic polarity because it lies where the negative and positive magnetic poles cancel each other out. We know from our simple key that black holes are giant magnets. When molecules of various materials come under the influence of magnetic fields they behave in one of two ways. They can be attracted towards the source of the magnetic field – in which case they are said to be 'paramagnetic' materials – or else they can be repelled away from the source of the magnetic field – in which case they are said to be 'diamagnetic' materials. It is this *para* pulling and *dia* pushing magnetic force that ultimately gives us the dark matter effect. We also have ferromagnetic materials that either have their own magnetic fields or generate one under the influence of another magnet and are thus obviously strongly paramagnetic. Any molecule on either side of a black hole will be influenced by either of its two magnetic fields, but molecules lying in the quiet space between its two magnetic poles, where the two magnetic fields cancel each other out, will not be affected. This quiet space can be seen in the middle of our photograph, lying in a circle between the positive and negative poles, and is where material will gather to form the discs of disc galaxies. So, all the attracted paramagnetic and ferromagnetic material approaching the black hole will be pulled in towards the centre and will be assimilated into the magnetic core as it grows ever larger. But all the repelled diamagnetic material will be pushed away from either side of the black hole's magnetic fields and will accumulate in the shape of a disc in the quiet space between the positive and negative magnetic fields. As these diamagnetic molecules are being pushed away into the disc, they are continuously being accelerated by the repelling force of the black hole, so when they reach the outer rim of the disc galaxy they are travelling much faster than would be expected. This is why the material in the outer edges of the disc galaxy's rim is travelling so fast and why scientists believe dark matter must exist to create this effect. Our simple key tells us that dark matter does not exist – it is the pushing effect of the black hole's magnetic fields repelling the diamagnetic material that gives the material in the disc this extra speed.

There is one more force to consider when creating the impression of the existence of dark matter. For the dark matter effect to take place, all of the accelerated material moving towards the disc from either pole of the black hole has to be moving in the same direction, in order to give circular momentum to all matter in the disc. This occurs due to what is known as Ampère's right-hand screw rule. Ampère's rule describes the twisting force exerted on charged particles when moving in a magnetic field. If it were not for Ampère's right-hand rule, the material would be forced into the disc with the random momentum it carried into the sphere of influence of the black hole, and the outer rim of the galaxy's disc would rotate as expected without the influence of any dark matter. If you close your right hand with thumb extended and apply it to what you see in a disc galaxy, your fingers will point in the direction of rotation of the matter in the galaxy's spiral arms and your thumb will point towards the negative magnetic pole of the black hole at the centre of the galaxy.

To summarise, our simple key tells us that dark matter does not exist, but we think it does because a black hole is a giant magnet made from all the magnetic quarks it has consumed while digesting paramagnetic and ferromagnetic material it draws in from space. The magnetic field of the black hole repels diamagnetic material away from each pole and out into the black hole's spiral galaxy disc, and by this process gives this material angular momentum, the direction being determined by Ampère's right-hand rule. It is this extra momentum of matter in the outer rim of the disc that leads scientists to believe that dark matter exists.

There is one final point of interest our simple key tells us about the nature of spiral galaxies. As carbon, oxygen and water are all strongly diamagnetic materials, they will have been repelled by our galaxy's black hole and been propelled into its spiral disc. This is why we find such an abundance of these materials in all of the planets and comets we explore in our solar system, something that has puzzled scientists in recent years, especially as oxygen and water are such delicate molecules and are so easily destroyed. This diamagnetic repelling process will also ensure an abundance of these basic building blocks of life in all of the other spiral galaxies on our side of the universe and of course the oppositely charged anti-matter building blocks for anti-life on the other side. You may ask why we also find ferromagnetic and paramagnetic materials in our galaxy's disc; after all they should have been consumed by our black hole. Remember that all of these elements are being created all of the time in the burning of billions of stars in our galaxy's disc that has formed in our black hole's magnetically neutral zone.

6. Quasars and the Anti-matter Boundary

Quasars are supermassive black holes that are thought to have formed out of the collisions of smaller galaxies. They emit enormous amounts of electromagnetic energy as they consume matter that is pulled into their centres under gravity. They appear in the sky as bright stars that shine with 100 times the intensity of all the light emitted by our whole galaxy. They are very distant and very ancient, around 12 billion light years old. As they consume matter, they release jets of subatomic particles funnelled along the axis of their swirling discs. Scientists observing these jets or beams of light from all of the quasars so far studied have discovered that they are all pointing in the same direction. They are puzzled by the fact that the jets run parallel to each other even though they are separated by billions of light years in space. How can such massive objects, separated by such vast distances, know in which direction to point their poles so that they are all aligned?

Our simple key suggests an answer for this phenomenon. Let us look again at our photograph representing the universe before the Big Bang.

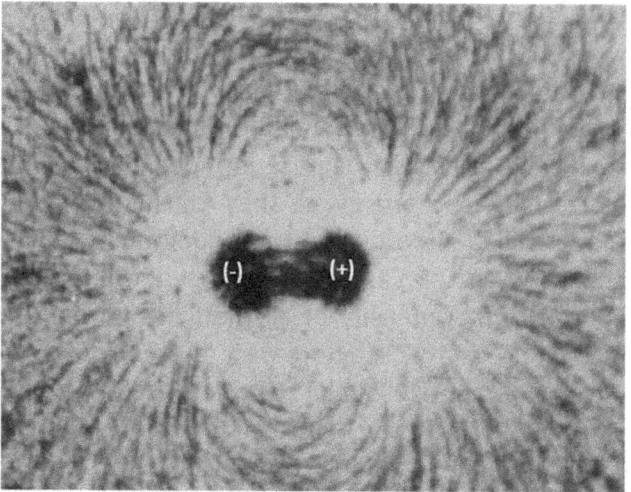

The most likely place that quasars will form in the universe will be in the magnetically quiet band between the positive and negative poles, the same place stars and planets form in disc galaxies. In the same way as black holes have a magnetically quiet band between their positive and negative poles, the universe has a quiet band between the positive and negative poles of the singularity that caused the Big Bang. Quasars will contain the quarks from the matter and anti-matter that have accumulated in the universe's magnetically quiet disc. When they form in this place, due to the influence of the universal magnetic fields left behind after the Big Bang, they will align their poles and jets of subatomic particles in the same direction as the singularity's remnant magnetic field lines. If the positions of all of these aligned quasars are plotted in time and space, they should fall along a single plane, the universe's disc, and will mark the boundary between matter and anti-matter.

7. Aether, the Michelson–Morley Experiment and Faraday's Paradox

The concept of 'aether' was used in several theories to explain natural phenomena, such as the how light travels through space. In the late 19th century, physicists postulated that aether permeated all throughout space, providing a medium through which light could travel in a vacuum. In 1887, two eminent scientists, Michelson and Morley, conducted an experiment to search for the aether and concluded that it did not exist. These aether theories are considered to be scientifically obsolete as Einstein's development of special relativity showed that Maxwell's equations do not require the aether. However, Einstein himself noted that his own model which replaced these theories could itself be thought of as a type of aether, as it implied that the empty space between objects had its own physical properties.

Our simple key provides the 'aether' in the form of the two universal magnetic fields, something Michelson and Morley were not looking for because they would at least have found the Earth's magnetic field. They were looking for something else to transmit light and explain gravity. The really big mystery today is how science still points to the Michelson–Morley experiment as proof that aether does not exist. Our simple key requires the aether in the form of a universal magnetic field to explain everything in our physical universe today and to make sense of the mathematical equations our scientists use as the basis for calculating the inner workings of molecules and electronic devices. We need the aether to have two halves, one positive and one negative to provide for the existence of matter and anti-matter. We need the aether on our side of the universe to provide the single magnetic charge that makes the quarks appear lopsided, in order that three of them can form a proton with its electron. We need the negatively charged aether to provide Dirac's electron sea, which forms the basis of our mathematical understanding of the quantum realm. We need the aether to be inversely squared in shape, to explain the dark energy push that is causing the accelerated expansion of the universe. We need a magnetic aether for the neutrons to tug on and stretch to create the holes in space-time that bodies fall into for our explanation of gravity. We need the aether to provide the magnetic field that gets tugged and pushed by fast spinning particles to make light and other wavelengths of electromagnetic radiation, and give them something to travel through so we can see them on the other side. We need the magnetically charged aether to keep electrons of the same charge bunched up or magnetically bottled up

in order that they can appear to travel through space
as a wave but arrive as a particle. It makes perfect
sense that the medium through which light and
electrons travel is made from the same stuff as they
are, after all, water waves travel through water. I am
not sure what Michelson and Morley were hoping to
find when searching for the medium that transmits
electromagnetic radiation through a vacuum. I like
the idea of electromagnetic waves travelling through
magnetic fields. The problem with our magnetic
field on our side of the universe is that it is mono-
polar, only negatively charged, and so we can't see
it for what it is. We are very much a part of it and a
product of its existence, from the very smallest atom
to the largest quasar.

Our simple key, with its inversely squared
universal magnetic field, also neatly explains
Faraday's Paradox. What Faraday found to be so
contradictory when experimenting with magnets
and electricity is basically as follows; when a coil
of wire is rotated around a stationary magnet, an
electrical current is generated within the wire and
when a magnet is rotated around a stationary coil of
wire, an electrical current is again generated in the
wire. However, if a magnet is rotated about its own
axis within a coil of wire, no current is generated.
Thinking about our inversely square shaped universal
magnetic field and indeed the fields of all magnetic

bodies, the shape of those fields will trap and hold the field of any magnet that is being rotated about its own axis. The magnetic field cannot move because it will be trying to move against the inversely squared slopes of all of the surrounding magnetic fields. It will be trapped and held by the slopes of the fields and will only be able to move in any direction if its generating magnet is moved.

Conclusion

I hope you have as much fun using this simple key to explain the physical world around you as I have had in developing it. I also hope you get the very rare opportunity to corner a particle physicist, astronomer or perhaps even an eleven-dimensional string theorist and explain simply to them how the universe really works. Remember; don't be frightened to use your imagination, for imagination is more important than knowledge. Knowledge is limited to all that we now know and understand, while imagination embraces the entire universe, and all there ever will be to know and understand.

www.ingramcontent.com/pod-product-compliance
Lightning Source LLC
Chambersburg PA
CBHW071125210326
41519CB00020B/6427